Bilingualism and Cognitive Ageing:

Does learning a new language help prevent cognitive ageing?

Bilingualism and Cognitive Ageing:

Does learning a new language help prevent cognitive ageing?

Written by
Dr. Austin Mardon
Catherine Mardon
Ashmita Mazumder
Jiyeon Park
Anna Yang
Camryn Kabir-Bahk
Haya Sonawala

Edited by
Taryn Foster
Kathryn Carson
Naiya Patel
Jeremy Steen

GM
PRESS

First Printing: 2021

Cover Design and typeset by Clare Dalton

ISBN 978-1-77369-616-4

E-book ISBN 978-1-77369-617-1

Golden Meteorite Press

103 11919 82 St NW

Edmonton, AB T5B 2W3

www.goldenmeteoritepress.com

Contents

Chapter 1: The history of research on the cognitive effects of bilingualism

Written by Ashmita Mazumder

Introduction

Until about 50 years ago, there was a general agreement in the field of cognitive psychology that bilingualism had negative effects on developing minds. The conversation on bilingualism and its effects began as early as 1913, when Ronjat published their article suggesting that more than half of the world's population was multilingual to some degree.[1] This report garnered significant attention and researchers started to become interested in the cognitive differences between those who spoke multiple languages in comparison to those who did not. However, the definition of bilingualism or multilingualism was not established at this point. While some researchers considered any "person who spoke two languages on a regular basis for most of their lives" as being bilingual,[2] others were more apprehensive and considered only those "who could switch effortlessly between two languages and adopt appropriate sociocultural stances for each of them" as bilingual.[2] Currently, bilingualism is not considered a dichotomy where one is either a bilingual or a monolingual.[3] Rather, bilingualism is now considered to be a continuum where each individual is placed based on their unique experience. In this chapter, we will delve into the history of research on bilingualism and the development of proper research methods for the same.

This chapter will highlight the changes that took place in the field that led to better research outcomes and fair comparisons between monolinguals and bilinguals.

Early Studies

Most of the research conducted between 1920-1960 reported bilingualism as a major issue. The main concern regarding bilingualism following the 1920s was the effect it would have on the intellectual abilities of children.[4] Many studies reported that bilingual children had a poorer vocabulary, deficient articulation, lower standards in written composition, and more grammatical errors. The results of these studies suggested that bilingual students suffered from some language handicap.[5] These consistent results encouraged researchers to investigate the effects of bilingualism on overall intelligence as well. At the same time, psychometric tests of intelligence were being developed which prompted researchers to think about the validity of these tests on bilinguals.[6] Since these tests predominantly depended on verbal abilities, the researchers were sure that any differences between bilinguals and monolinguals would be highlighted through the results of these tests.

Following this, multiple studies were published which reported the underperformance of bilingual students on intelligence tests. Researchers believed that learning or knowing two languages could lead to confusion and expected to see signs of mental retardation in children who learned to speak more than one language.[6] One such study published by Saer[4] in 1923 used the Stanford-Binet intelligence scale to find the relationship between bilingualism and intelligence.[4] The author recruited children aged 7 years and tested them annually. The study reported that rural bilingual children had significantly lower scores on the test compared to rural monolingual children.[4] More interestingly, Saer found that with each year, bilingual children seemed to score lower on the test than the previous score.[4] He concluded that they were facing "mental confusion". Pintner in 1932 compared results from non-language and language tests in an English-speaking and a non-English speaking group.[7] The non-English speaking group contained participants that spoke a language other than English at home. Contrary to Saer's results, Pintner did not find any differences between monolingual and bilingual students.[7] Pintner's study sparked a conversation on the effectiveness of the study design being used by researchers to date.[7] As intelligence tests being developed and used by researchers at the time tested verbal abilities only, many were concerned about whether these scores reflected overall intelligence.[6] Some researchers also found it unfair to judge and rank those whose primary language was not English based on the score they obtained on tests that were developed for English-speaking populations.[6]

Moreover, concerns were raised about the validity of the experimental design. Researchers did not make enough efforts to control extraneous variables that may affect measures of cognitive ability. These studies did not match participants along several dimensions such as socioeconomic status (SES), gender, and age. In 1930, McCarthy noted that bilingualism in the United States was correlated with low SES and that more than half the occurrences of bilingualism in school were classified as belonging to families from the unskilled labor occupational group.[5] At the same time, Fukuda[8] (1925) reported in their paper that most of the high-scoring English-speaking subjects belonged to the occupational and executive classes and even reported a correlation of 0.53 between the Whittier (socioeconomic) scale and the Binet IQ for this population.[8] Under this observation, it was unfair to compare the intellectual abilities of those from the low SES families to those in the high SES families. A second major concern was the operationalization of the degree of bilingualism in participants.[10] Some papers merely determined the degree of bilingualism by the foreignness of parents which was highly problematic as it did not give any information about how comfortable the participant was in speaking two languages and being able to switch between them. In other studies, the sample's bilingualism was assessed through family names or place of residence.[9]

Defining Bilingualism

Starting in 1937, Arsenian published an article arguing against the unidimensional concept of bilingualism.[10] They argued that variations between bilingual experiences could make a huge difference in the type of effect it has on a sample's cognitive performance.[10] He proposed specific dimensions along which bilingual samples could be defined.

Degree of Bilingualism[10] _

Bilinguals differ in degree of proficiency in their two languages. While some children are just beginners in learning a second language, others have achieved age-appropriate levels of proficiency in both languages. The effects of such variations should be taken into account when considering the abilities of bilinguals.

Degree of Difference between the two Languages

The degree to which two languages differ is also an important factor to consider when investigating the degree of bilingualism. While Spanish is closer to other European languages such as Italian or French, it is very different from English or Japanese. Therefore, more cognitive effort would be required from a Spanish child to learn English than to learn French. Furthermore, languages also represent deeper cultural differences that a child must accommodate.

Age when Learning the Second Language

The experience of infants exposed to two languages is different from the experience of a monolingual 6 or 7 year old trying to learn a second language and a balanced school curriculum. This difference in experience should be accommodated when working with bilingual samples.

Method of Learning the Second Language

Researchers should document whether the participant had learned two languages simultaneously or whether the second language followed after completely acquiring the first language. In the first case, the child has "acquired" the second language which refers to the process of acquiring a second language in a natural process and an environment outside of formal instruction. In the second case where the second language followed the first, the child made a conscious attempt to "learn" the language. In this case, each aspect of the language is formally introduced and feedback and correction is provided.[11]

Attitudes toward the Second Language

Experiences of bilingual samples can vary in terms of social, political, and religious sentiments related to the first and second languages learned. Having to learn a second language might threaten a person's self-esteem, especially if the language is identified in any way with a colonizing or assimilating force.

Arsenian believed the five factors above should be accounted for in studies as they present a clearer picture of the sample and their degree of bilingualism. However, most studies before 1962 did not consider these factors at all.

In the latter part of the 20th century, researchers began playing individuals into categories depending on various socio-cultural factors and the participants' language one and language two proficiency. Cummins presented the 'threshold hypothesis' which stated that "a minimum threshold in language proficiency must be passed before a second-language speaker can reap any benefits from language".[12] This hypothesis was widely recognized and used as a device in bilingual studies.

Cummins' Threshold Hypothesis divided bilinguals into three categories[12]. Balanced bilinguals are regarded as individuals that have achieved high competence in both language one and language two. Dominant bilinguals prefer one language over the other and therefore, develop more competence in one of their languages. Semilinguals are thought to be individuals that have lost a significant amount of competence in both language one and language two.

The Shift

A breakthrough study in the field of bilingualism was published in 1962 by authors Peal and Lambert.[13] Initially, the authors expected to find results consistent with previous findings where monolinguals would score higher than bilinguals in intelligence tests.[13] However, these authors implemented a methodological approach in which they matched bilingual and monolingual participants in terms of their socioeconomic status, language proficiency, language experience, gender, and age.[13] This improvement in the experimental design ensured that the results were not confounded by extraneous variables. The researcher reported that contrary to their predictions, bilinguals outperform monolinguals on several measures of verbal and nonverbal intelligence.[13,14] The authors explained this unexpected result by suggesting that bilinguals exhibit mental flexibility and present a diversified set of mental abilities.[21]

The 1990s witnessed a rise in the number of articles published on bilingual research. Inspired by Peal and Lambert, researchers started looking at bilingual children to replicate their findings. The benefits of bilingualism were brought to the forefront through the consistent results that researchers found in this period.[2] However, the evidence of bilingual children outperforming their monolingual counterparts accumulated following the shift from examining the effect of bilingualism on language-related tasks to a focus on cognition.[8] Bilingual children showed a heightened ability to solve linguistic problems.[16] based on metalinguistic awareness, bilinguals are able to represent language as a process and an artifact rather than just words.[17] They are able to solve required nonverbal problems by inhibiting misleading information.[18] Researchers started to design studies to parse the exact processes by which bilingual children performed better on meta-linguistic tasks.[16,19] Bialystok and colleagues suggested that it had less to do with the linguistic knowledge and more to do with the enhanced capacity to selectively attend to competing cues or constraints.[17] Following this, in 2012, Miyake and Friedman's tripartite theory stated that there are three aspects of executive functions: updating, inhibition, and shifting.[2] Under this framework, updating is defined as the continuous monitoring and quick modification of contents within the working memory,[2] inhibition as the suppression of prepotent responses, and shifting as a change in mentality.[20]

Recent studies have also reported that bilingualism may offset some age-related cognitive declines. Bialystok and colleagues found that bilingual adults developed symptoms of dementia four years later than monolinguals even when all other factors remained constant.[21,22] They state in their paper that "the lifelong experience of managing two languages attenuates the age-related decline in the efficiency of inhibitory processing".[3] Therefore, the bilingual advantage was seen as a marker of enhanced executive functioning, the processes involved with mental control and self-regulation to achieve a goal.[23] Bilinguals can selectively attend to different representations which may be responsible for their greater attentional control.[23]

Conclusion

In the last two decades, many studies have presented the different effects that bilingualism has on an individual's cognitive and linguistic abilities. Early reviews on bilingualism had poorer control conditions and did not implement experimental designs which allowed a fair judgment of the intellectual capacities of bilingual individuals. However, after several attempts were made to properly define bilingualism and articles were published on how to categorize individuals based on their degree of bilingualism, the field witnessed a shift. With better experimental controls and designs, researchers were able to uncover the bilingual advantage. Bilinguals were reported to consistently perform better than monolinguals in verbal and nonverbal intelligence tests. Multiple tests also report slower cognitive aging and better attentional control.[23] Bilinguals were reported to have better executive functioning and mental flexibility.[21] The field of bilingual research is an example of the effects that poor experimental design can have on the results of a study. Proper implementation and statistical controls are needed to avoid the effects of extraneous variables on the outcome variables.

Chapter 2: Ages and bilingualism

Written by Jiyeon Park

Introduction

Aging is a natural and inevitable part of life that all living
creatures experience as time passes. It becomes apparent in social,
behavioural, psychological, environmental, physiological and
biological processes.[1] Since human beings are social animals, they
are eager to communicate with each other.[2] Language therefore
plays a vital role as it allows individuals to express their own
opinions and thoughts, both orally and in writing. However,
communication skills and the use of language may change subtly
in the aging process since aging is associated with physiological
changes in speech, voice, and hearing. Because of these changes,
the ability to learning new skills and acquiring another language
may be slightly challenging to certain groups of people, such
as seniors, or people with dementia or Alzheimer's disease.[3] In
this chapter, we will learn the optimal ages to learn a language
and discuss the impact of learning a language on one's cognitive
abilities. We will additionally take a closer look into the concept
of bilingualism and the three types of bilingualism.

When to Learn a Language

Each individual is unique and has different strengths. As a result, each person has a slightly different period in life that is optimal for language learning. As social beings, communication is key for human beings and therefore, the need and desire to communicate stimulates both the use and development of language. This desire, of course, becomes stronger when an individual has someone with whom to communicate.[4] With the development of technology and easy access to others around the world, people's interest in learning a foreign language has increased. They value this skill for their resume, as well as a tool that will facilitate travel. Most countries already incorporate second language learning in the curriculum, but people's aspiration to learn a foreign language has been steadily gaining popularity.[5] So, when should individuals start learning a new language?

Without any instruction, young children learn their first language naturally as they grow up, through loving interactions and relationships with their caregivers. Then, what about learning a new, second language? Children who are raised by bilingual parents are exposed to an environment in which parents value both languages and wish to raise bilingual children. These children consistently hear two languages spoken by their parents and absorb the languages instinctively. Interestingly, many parents in the United States hire caregivers who can speak additional languages in order to teach the new language to their children before they enter elementary school.[6] On the other hand, some parents who raise children in a bilingual household are often concerned about confusing their children, mostly because of "code mixing". Code mixing is a term used to describe the phenomenon of bilingual children mixing and using words from two different languages in the same sentence; yet, this is a perfectly normal part of bilingual development. Since these children hear two different languages from adults around them, their vocabulary in one language may be limited, so they end up code mixing to communicate. However, this is a concern that is solved as time passes, so if parents witness code mixing in their children, they should take this as a sign of their bilingual child's ingenuity. In fact, code mixing has important and positive social implications as code mixing is a significant part of being part of a bilingual community such as Spanish-English communities in the United States and Afrikaans-English regions of South Africa. Therefore, it is not necessary for parents to intentionally discourage their bilingual children from mixing their languages.[7]

Notably, unlike bilingual children, bilingual infants can tell the difference between their two languages without confusion, because infants are sensitive to perceptual differences in languages and are attuned to the rhythm of a language. This allows them to distinguish rhythmically dissimilar languages such as French and English at birth. Later, by the age of four months, they can discriminate between rhythmically similar languages like Spanish and French. However, as infants age to eight months, there is some distinction between bilingual infants and monolingual infants because monolinguals begin to stop paying attention to differences between languages, whereas bilingual infants can continuously distinguish between their languages based on their sensitivity to information.[7] Moreover, second language learning is a much more challenging process for adults, than children, as adults require more effort and time to learn and master a new language.[5] Since human brains are more receptive to language at an early age and the environment is more conducive to language learning at the early onset of life, it is therefore ideal to start learning a new language at an early age. This is why many young children are able to develop their language skills quickly, earlier in life. However, age should not hold back an individual from learning a new language. As long as you are willing to learn and have the motivation to face the challenges, it is never too late to acquire a new language.[7]

The Definition of Bilingualism and the Three Types

Numerous children in North America and across the world are exposed to two languages as they grow up and consequently become proficient in two languages. These children are called "bilingual" children, as the term refers to the ability to use and speak two languages fluently in daily life. As there are many bilingual parents in the world, bilingualism has become a common part of this world and in fact, surprisingly, one in three people is either bilingual or multilingual. This phenomenon typically bridges between two continents such as Asia, North America, Europe and Africa.[7] As a result, the outcome of extra-linguistic skills is closely associated to cultural, political and economic relations. Bilingualism is thus a valuable aptitude that overrides territorial borders and forms social links between people.[8]

Bilingualism can be categorized into three different types, namely "compound," "coordinate," and "sub-coordinate." If someone learns two languages in the same environment, then this bilingual is referred to as a compound bilingual. For example, if a two-year-old child is raised by an English mother who mostly speaks English and an Italian father who uses both languages, then this child is considered to be a compound bilingual.[9] Since compound bilinguals acquire two verbal languages as one notion, typically an absence of independent grammar for the second language is seen among these bilinguals. Therefore, it is especially important for such children to learn a second language in a way that they can recognize each language independently.[10] A second group, coordinate bilinguals, are individuals who learn two languages in two different contexts. Therefore, they can independently separate the words of the two distinct languages. If two siblings aged five and nine always attend English-language schools but grow up under Italian parents, then these siblings are considered coordinate bilinguals.[9] Thus, unlike compound bilinguals, coordinate bilinguals are capable of distinguishing the distinct grammar of two languages, independently. However, they usually have difficulty in translating the two languages at the beginning as they have acquired two languages separately, but eventually overcome this as they age.[10] Lastly, if one language dominates over another, this is called sub-coordinate bilingualism. For instance, if teenage girls grow up under an English mother and have always studied at English-language schools, but are also studying French, then they are more proficient in one language than the other.[9]

The Significance of Bilingualism

Currently, there are many bilinguals residing in New York, California, Arizona, Florida, New Mexico and Texas and this number is continuously increasing. For instance, it is expected that over 50% of the children in California will be capable of speaking another language other than English by 2035. This trend is also found in some urban areas in Canada, including

Toronto, where up to 50% of students' native language is non-English.[7] Interestingly, many people believe that growing up in a monolingual, mainstream culture is one of the most natural ways to grow up, but in fact, monolingualism is far from the norm. This is because many immigrants who have moved to North America from another country for a better life and/or education highly value the ability to speak more than one language. Furthermore, they believe that the ability to speak a second language at a young age will lead children to greater affluence in the future. This is why parents around the world expose their children to new environments where they can learn to speak additional languages at an early age.[6]

What other benefits come from being bilingual? Popular articles and books such as "The Power of the Bilingual Brain" in *Time Magazine* and *The Bilingual Edge*, highlight the potential benefits of being bilingual at an early age. Some of the greatest advantages of early bilingualism are advancement in employment, ease of travel, social communication with extended family, and a connection to family history and culture. Bilingualism additionally provides opportunities to get to know people from different backgrounds and cultures and understand the thoughts and feelings of foreigners. For instance, bilingual preschoolers are noted for better skills in understanding others' thoughts, intentions, perspectives and desires, when compared to monolinguals.[7] Proficiency in multiple languages will also prove to be an extremely valuable attribute as globalization becomes central to our lives. Language facility will help to ensure one's success when starting a new life in a new country or needing to communicate in an academic area, professionally.[5]

The Age of Language Learning Affects Cognitive Function

One of the most important cognitive functions in everyday life is verbal communication. This is why it is essential to learn and implement words and the rules of language.[5] As previously mentioned, learning a new language early on is more advantageous. The benefits of being bilingual are both practical and physiological. So, how does learning a new language at a specific age affect cognitive function?

Through their language studies, researchers have recognized unique characteristics in young bilingual children that are not present in monolingual children. For instance, when preschoolers are having a conversation, they usually rely on what speakers say instead of interpreting and considering a speaker's emotion, which is the opposite of what adults do, who rely on tone of voice when communicating. Children usually have difficulty concentrating on a speaker's tone of voice as it is challenging for them to constrain their tendency to focus on content. However, numerous studies have revealed an improved and well-developed inhibitory control in bilingual children which offers them distinct advantages. For instance, such children can judge a speaker's emotions by listening to their sentences and thereby distinguishing whether it is a happy or sad voice.[11] Additionally, several studies demonstrate that bilingual children have greater flexibility performing cognitive tasks.[12] Researchers conducted comparative studies on monolingual and bilingual six-year-olds to examine their interpretations of four ambiguous images. The results highlighted that bilingual children were successful at finding other meanings in the images that monolingual children were not able to detect. This finding suggests that bilinguals are better at perceiving additional images, by inhibiting attention to a previous image. They are also excellent at solving problems by using this inhibitory control, thus achieving better performance than monolinguals.[13] Unlike monolinguals, bilinguals are better at performing tasks that require alternating between activities and inhibiting the responses that they previously learned. These cognitive advantages have been documented among all bilingual infants, toddlers, children and adults.[7] Another cognitive advantage for bilingual children is memory generalization during infancy. Since bilingual infants develop multiple language systems at an early age, this early development may affect domain-general processes like memory that maximize the bilingual's capacity to learn.[14] Furthermore, numerous studies on bilingualism and aging suggest that learning a new language in adulthood helps to avoid and/or delay decline in cognitive function. This is because bilingual people are used to adapting to an ever-changing environment in which they must constantly make accommodations to succeed. As a result, bilinguals become more mentally flexible and their cognitive functions are strengthened.[15] In short, being a bilingual definitely offers benefits to all age groups, ranging from infancy to school-aged children, due to an enhanced inhibitory control that monolingual children do not demonstrate, as well as other cognitive skills such as keen attention.[12,16]

Conclusion

Studies indicate that second language learning aids an individual in many ways. There is also much evidence that highlights how much easier it is to learn a new language at an early age. Researchers have identified three different types of bilingualism, each with their own characteristics, but overall, all advantageous to an individual's future.[9] Being a bilingual helps with communication with others in the world, as well as for future employment. Significantly, however, it also enhances cognitive functions such as inhibitory control, memory, and attention.[7] The mechanisms behind bilingualism as they affect the brain and the science behind bilingualism and cognitive aging will be discussed in later chapters.

Chapter 3: How bilingualism affects the brain

Written by Anna Yang

Introduction

There is an abundance of evidence that suggests that bilingualism affects the brain not only concerning cognitive aging but during development and throughout adulthood as well. For much of scientific history, this evidence primarily concerned the psychological effects of bilingualism and its effects on certain cognitive functions, but over the past few decades, technological advances have allowed researchers to peer deeper into the brain to investigate the physiological effects of bilingualism on the human brain. This chapter will discuss both the physiological and psychological effects of bilingualism on the brain and the mechanisms that are believed to underlie these impacts.

Physiological Effects of Bilingualism

It has been firmly established that bilingualism does, indeed, affect the structure of the brain in adults, as evidenced by the experience-dependent grey and white matter changes that have been observed in brain structures implicated in functions such as language learning, processing, and control.[1] In addition to mere structural changes, it has also been found that bilingual and monolingual people have distinct developmental trajectories in both grey and white matter structures, suggesting that there is a developmental basis to the well-documented structural differences observed between bilingual and monolingual brains.[1] In summary, the current evidence suggests that in adults, bilingualism-induced neuroplasticity cycles dynamically through several stages.[1] First, relatively early increases occur in regional cortical grey matter, likely due to the initial stages of language learning, use, and control.[1] The next stage consists of cortical grey matter renormalization, subcortical grey matter increases, and increases in white matter integrity as one becomes progressively more efficient in the learning, use, and control of the language in question.[1] This section will examine the physiological effects of bilingualism, namely on the structure and developmental trajectories of the human brain, in greater detail.

Grey Matter

An increasing number of studies exist which suggest that bilingual or multilingual adults show structural alterations in cortical regions and subcortical grey matter structures in comparison to monolinguals. These structural changes are generally observed in brain regions that are involved in bilingualism, which entails the learning of the knowledge and skills involved in the use of language (e.g. phonology, lexico-semantics, grammar) as well as the control between languages.[1] As such, bilingualism is expected to affect the structure of cortical and subcortical regions involved in language learning, processing, and control. Indeed, the reported structural effects of bilingualism in adults are most commonly reported in grey matter regions that have been found to underlie such language-related processes.[1] These regions include the frontal cortex (particularly the three portions of the left inferior frontal gyrus, namely the opercularis, triangularis, and orbitalis, the frontal pole, the middle and superior frontal gyri, and the anterior cingulate cortex), the temporal cortex (particularly the superior, middle, and inferior temporal gyri, Heschl's gyrus, the temporal pole, and the hippocampus), and the parietal cortex (particularly the supramarginal gyrus, the angular gyrus, and the superior parietal lobule).[1] The subcortical structures affected include the basal ganglia (in particular the caudate nucleus, the putamen, and the globus pallidus), as well as the thalamus, with some effects also having been reported in the cerebellum.[1]

The specific structural effects of bilingualism in adults appear to vary as a function of both experience and age.[1] For instance, in young adult bilinguals with limited experience using their second language, studies have found cortical tissue increases reflected as increased cortical thickness and/or cortical volumes.[1] These increases have been observed in multiple frontal, temporal, and parietal regions in bilinguals.[1] However, young adult bilinguals with limited experience using their second language often show limited or no changes in subcortical volumes, whereas subcortical differences have been observed in more experienced bilinguals, particularly in those with substantial immersion in second language speaking environments.[1] These bilinguals show greater volumes than monolinguals in a variety of subcortical structures including the basal ganglia, thalamus, and cerebellum.[1]

In contrast, experienced bilingual young adults generally do not show differences in cortical thickness or volume in comparison to monolinguals.[1] Somewhat different patterns have been observed in older bilingual adults, who generally show greater grey matter volumes in these cortical and subcortical regions, suggesting that both experience and age are important contributing factors to the physiological effects of bilingualism.[1] The greater grey matter cortical and subcortical thickness and volume observed in older adult bilinguals compared to older monolinguals is generally interpreted as the result of greater age-related tissue loss for monolinguals.[1] It has also been suggested that at least some of the grey matter increases observed in adult bilinguals relative to monolinguals can be explained by a lower amount of pruning in bilinguals during brain development.[1]

Concerning the developmental trajectory of grey matter, there have been significant differences observed between bilinguals and monolinguals as well. Bilinguals tend to show less developmental loss (i.e. more grey matter) beginning in late childhood and adolescence, relative to monolinguals.[1] These developmental differences are observed primarily in the frontal and parietal regions of the brain, particularly in the inferior frontal gyrus pars opercularis, superior frontal cortex, inferior and superior parietal cortices, and precuneus.[1] There are currently several models that attempt to explain the effects of bilingualism on the developmental trajectory of grey matter. Drawing from these models and the common denominators that they share, Christos Pliatsikas proposed the dynamic restructuring model (DRM) of structural changes in the brain due to bilingualism.[1] The DRM is based on the expansion-partial renormalization hypothesis (EPH), which states that the acquisition of a new skill is marked by transient local expansions in the involved grey matter regions, manifesting as increases in cortical thickness or volume.[1] These increases in grey matter tissue are believed to be due to the development of new dendritic spines, and possibly neurogenesis as well.[1] According to the EPH, these increases are followed by a slow pruning process where older and idle spines are removed, eventually resulting in the reversal of the initial expansions as the newly formed networks become more efficient with time and practice.[1] Putting the EPH into the context of bilingualism, Pliatsikas suggested that the local grey matter increases reported in relatively inexperienced bilinguals reflect an initial tissue increase in brain structures related to language learning and use.[1] However, as language use becomes more efficient with increased experience and practice, these structures gradually return to baseline.[1] This explains the patterns observed in bilinguals, where cortical thickness and volume increases are seen in less experienced bilinguals relative to monolinguals but no such differences are evident in more experienced bilinguals.[1] It is worth noting, however, that larger subcortical volumes are observed in experienced bilinguals relative to monolinguals.[1] These differences are thought to be explained by the involvement of these subcortical regions, particularly the basal ganglia, in language control as well as later stages of language learning.[1]

White Matter

Bilingualism is also associated with structural changes in several white matter tracts that provide connectivity between the cortical and subcortical structures underlying language-related processes.[1] These include the inferior and superior longitudinal fasciculi, the inferior fronto-occipital fasciculus, the arcuate and uncinate fasciculi, the anterior thalamic radiation, and the corpus callosum, including the forceps major and forceps minor.[1] Like with grey matter, the white matter effects of bilingualism appear to be experience-dependent.[1] For instance, young adults with limited second language experience generally do not demonstrate white matter effects, whereas experienced bilinguals demonstrate increases in white matter integrity.[1]

Neurobiologically, increases in white matter integrity have been associated with greater myelination in the implicated tracts.[1] Axons that are more active are expected to be associated with greater myelination, which may, in turn, be observed as greater white matter integrity.[1] As such, white matter integrity is at least partly a function of experience and can be altered due to the acquisition and/or use of a new skill, such as a new language. Putting this into the context of bilingualism, Pliatsikas argued as part of the DRM that increases in white matter integrity emerge in more experienced young adult bilinguals, as has been observed, due to increased optimization of processes such as proceduralization and various aspects of language use and control.[1]

With respect to developmental trajectories, bilinguals have higher white matter integrity compared to monolinguals starting in mid-late adolescence, specifically in the striatal-inferior frontal fibers.[1] Lower fractional anisotropy values are observed in bilinguals than monolinguals starting from age 3, with the groups gradually converging until the bilinguals begin to show larger values again starting from age 16.[1] Overall, in both bilinguals and monolinguals, white matter integrity increased continuously with age, demonstrating the patterns typically expected of brain development in both cases.[1] However, this suggests that there is a greater increase for bilinguals than monolinguals in white matter integrity throughout development.[1]

Psychological Effects of Bilingualism

Attentional Control

As remarked by Ellen Bialystok, a professor of psychology at York University who studies the effect of bilingualism on cognitive function, "the cognitive functions that have been shown to be impacted by bilingualism largely concern attention - the ability to focus attention on relevant information and shift attention as needed."[2] This ability, known as attentional control, is one of the most central aspects of cognitive function throughout one's life, and it has been demonstrated that bilinguals tend to perform better in this area than monolinguals. Researchers have shown that the enhanced ability of the bilingual brain when it comes to attention and task-switching likely stems from its developed ability to inhibit one language while using another.[3] In monolingual people, areas in the frontal and temporal language regions, namely the left supramarginal gyrus and left inferior frontal gyrus, are activated when faced with phonological competition - when phonemes from different words compete with each other during phenological encoding.[2] Studies have shown that different areas of the brain are needed to cope with phonological competition from within the same language, as monolinguals do, compared to the between-language competition, which is what bilinguals frequently face.[2] As a result, the size, and type of the neural network that bilinguals recruit to resolve phonological competition differ depending on the source of competition.[2] When competition occurs between two languages, additional frontal control, and subcortical regions, specifically the right middle frontal gyrus, superior frontal gyrus, caudae, and putamen are recruited.[2] Increased use of these additional brain regions is believed to be the underlying cause of bilinguals' enhanced attentional control relative to monolinguals.

Task Switching

Another skill that is often considered an indicator of cognitive functioning is task switching - the ability to switch from one task to the next. Bialystok and her team published a paper in the journal *Cognition* sharing the results of an investigation into bilinguals' task-switching ability.[2] According to first study author Dr. John Grundy, the "experience of bilingual infants that requires them to pay attention to multiple sources of input within various linguistic contexts makes it adaptive for them to rapidly disengage attention from stimuli once they are processed so that attention can be re-engaged to currently relevant stimuli."[2] As a result, bilinguals tend to be more adept at task-switching than monolinguals. In a series of three studies involving a total of 145 bilinguals and 126 monolinguals, participants completed a test to study their ability to switch between types of stimulus displays where different responses were required.[2] The results show that bilinguals were faster at disengaging their attention from one trial so that they could focus on the next trial when a different response was required.[2] For instance, when switching from categorizing objects by colour to categorizing them by shape, bilinguals were able to do so more rapidly than monolinguals, reflecting better cognitive control when changing strategies on the fly.[2]

Language Co-Activation

Another psychological effect of bilingualism on the brain is language co-activation, which is the phenomenon observed in bilinguals where both languages are simultaneously active even when the person is only using one of them.[3] When one hears a word, the entire word is not heard all at once. Instead, the sounds arrive in sequential order, and as a result, the brain's language processing system begins to guess what the word might be long before the word is complete.[3] This guessing process involves the activation of multiple words which match the signal received.[3] For instance, if one hears "can," words such as candy and candle will be activated. Language co-activation refers to the observation that in bilinguals, this activation is not limited to a single language; instead, auditory input activates corresponding words across all of the languages known by the person in question.[3]

Some of the most compelling evidence that currently exists in support of language co-activation comes from studies of eye movements. People tend to look at the things that they are thinking, talking, or hearing about.[3] As such, eye movements serve as indicators of what people are thinking about at a given moment in time and have demonstrated the existence of language co-activation among bilinguals. For instance, it has been observed that Russian-English bilinguals asked to "pick up a marker" from a set of objects tend to look more at a stamp in that collection of objects than someone who does not know Russian, since the Russian word for stamp, "marka," sounds like the English word "marker."[3]

Inhibitory Control and Conflict Management

Bilingualism generates not only cognitive benefits, but certain language difficulties as well. For instance, having to deal with persistent linguistic competition across multiple languages can cause bilinguals to name pictures more slowly and increase tip-of-the-tongue states, where one is unable to fully conjure a word but can still remember specific details about it.[3] As a result, being able to juggle multiple languages creates a need to control how much one accesses a language at any given time. To maintain this balance between languages, the bilingual brain relies on executive functions, such as attention and inhibition, to a greater extent than the monolingual brain.[3] Because both of a bilingual person's language systems are always active and competing, that person uses these control mechanisms every time they speak or listen. This constant practice strengthens these functions, resulting in bilinguals often performing better on tasks that require conflict management. For instance, in the classic Stroop task, where people see a word and are asked to name the colour of the word's font, people are generally able to correctly name the colour more quickly when the colour and word match (i.e., the word "red" printed in red) than when they do not match (i.e., the word "red" printed in blue).[3] Bilinguals, however, typically perform better on this task than monolinguals, due to the task's involvement of one's inhibitory control ability.[3] When the word itself conflicts with its font colour, the brain must employ additional resources to ignore the irrelevant word and focus on the relevant colour. This ability to ignore competing for perceptual information and focus on the relevant aspects of the input is known as inhibitory control and is exercised much more frequently and consequently more effectively by bilinguals than monolinguals.[3]

Auditory Attention

The psychological effects of bilingualism extend beyond cognitive functions associated with the cortex to sensory processing, which is traditionally associated with the subcortical brain areas.[3] When monolingual and bilingual adolescents listen to simple speech sounds such as the syllable "da" without any intervening background noise, they show highly similar brain stem responses.[3] However, the researchers play the same sound to both groups in the presence of background noise, the bilinguals' neural responses are considerably larger, reflecting better encoding of the sound's fundamental frequency, a feature closely related to pitch perception.[3] This boost in sound encoding appears to be related to advantages in auditory attention. As articulated by Dr. Viorica Marian, "the bilingual juggles linguistic input and, it appears, automatically pays greater attention to relevant versus irrelevant sounds."[4]

Conclusion

Evidence accumulated from studies conducted over the last few decades provides substantial support for the assertion that bilingualism has both psychological and physiological effects on the brain. These effects are not confined to the later periods of one's life, during cognitive aging; instead, evidence suggests that the effects begin early in one's life, impacting the developmental trajectory of the brain and continuing throughout adulthood.

Chapter 4: Influential studies and scientists

Written by Camryn Kabir-Bahk

Introduction

The correlation that exists between cognition and bilingualism has been questioned and investigated for many years. For a very long period of time, the vast majority of people felt that bilingualism had very negative impacts on one's cognitive ability.[3] Scientists and researchers, specifically in the first half of the twentieth century, published information that argued that bilingualism resulted in a person being "mentally confused" or less competent compared to their monolingual counterparts.[3] Throughout the twentieth century, society began to shift its opinion on the relation between cognitive abilities and bilingualism. Interestingly enough, nowadays, people tend to feel that bilingual and multilingual people are actually much more intelligent than monolingual people. In addition to the societal shift regarding bilingualism, scientists began to think about how bilingualism might have an effect on cognitive ageing since it seems to affect cognitive abilities positively. In fact, there are many studies much more recently that have shown that bilingualism has a positive effect on cognitive ageing.[1] This chapter will focus on the significant studies that have been conducted that have demonstrated the correlation between slower cognitive ageing and bilingualism and the influential scientists that have helped to secure these major breakthroughs.

Lothian Birth Cohort

In 2014, a fascinating study was published in the journal *Annals of Neurology* which is one of the first studies to be conducted that studied the effects of bilingualism on cognition in older adults.[1] The study was conducted in two main waves.

Participants and Method of Selection for Participants in the Study

The first wave was done in 1947 and tested 1,091 participants who were 11 years old at the time.[1] The participants were selected from the Scottish Mental Survey, which looked at children born in 1936 and who were attending Scottish schools in 1947.[2] The objective of the Scottish Mental Survey was to assess the average intelligence of children born in 1936 and going to a Scottish school. The second wave of the Lothian Birth Cohort study was conducted between 2008 and 2010.[1] During the second wave of the study, 866 participants from the initial wave returned.[1] Of those 866 participants, 410 of the participants were female, and 443 were male.[1] Additionally, the mean average age of the participants in the second wave of the study was 72.49.[1] The participants varied slightly in their ethnic background because thirteen of the participants were of British descent and moved to Scotland before turning 11 years old.[1] The remaining 1,078 participants were all born in Scotland are of Scottish descent. However, the results from the study concluded very minimal differences between the cognitive performance of participants born in Scotland compared to Great Britain because the analysis for the study was conducted twice, once with the British descent participants and once without.[1]

Since the Lothian Birth Cohort study was investigating bilingualism and its correlation with cognitive ageing, the participants for the study also had to satisfy the condition of bilingualism. To do this, possible participants partook in a questionnaire.[1] However, the questionnaire was conducted during the second wave to ensure that participants who became bilingual after the age of 11 were still accounted for within the study. The questionnaire included the following questions:

1.   Do you know any other languages besides English?

2.   How many languages do you know?

3.   At what age did you learn these additional languages?

4.   How often do you utilize the additional languages you know?[1]

The fourth question was further broken down into how often the participants used their additional languages daily, weekly, monthly, or never using their additional languages.[1] The last question also asked participants about which domain they use the additional languages they know; in conversation, literature or media.[1] Any participants that said "yes" to the first question were categorized as bilingual and satisfied the bilingual condition to partake in the study.[1]

Methods Used to Evaluate Cognition

Once the participants were selected for the study, their intelligence had to be tested. However, since the participants for the study were all being selected from the Scottish Mental Survey, which was investigating childhood intelligence, the data collected from that study was used for the bilingualism study. After this, the first wave of the study was essentially completed. The second wave of the study then occurs between 2008 and 2010. During this time, the cognitive abilities of the participants are reassessed. The Lothian Birth Cohort study used six tests to investigate cognition: general fluid-type intelligence (g-Factor), memory, speed of information processing, Moray House Test, vocabulary and reading, and verbal fluency.[1] To investigate general intelligence and comprehension, general fluid-type intelligence was evaluated. General fluid-type intelligence is an excellent indication of how someone is performing cognitively because fluid intelligence focuses on how someone is able to process and think abstractly, reason, and logically solve problems.[4] Fluid intelligence was tested using six nonverbal tests coming from the different subtests of the Wechsler Adult Intelligence Scale-III, UK Edition and the Wechsler Memory Scale-III (WMS-III), UK Edition.[1] The Wechsler Adult Intelligence Scale is an intelligence quotient test that aims to test the IQ of adults and adolescent children.[5] The test was first developed in 1955. However, the version of the test used in the study is the Wechsler Adult Intelligence Scale-III (WAIS-III), which was developed and published in 1997.[5] The test uses various subtests to provide number scores on and help gauge how a participant does with speed processing, verbal comprehension, working memory and perceptual reasoning.[5] The average number score that is received on the tests is 100; however, about 66% of people who take the test will receive an average score that falls between 85 and 115.[5] The Wechsler Memory Scale-III was designed to look at different functions related to memory within a person; specifically, auditory memory (delayed or immediate), visual memory (delayed or immediate), working memory, general memory, and delayed auditory recognition.[7] One of the six nonverbal tests the researchers used was the Letter-Number Sequencing subtest to evaluate an individual's short-term memory.[1] The Matrix Reasoning subtest was used to examine visual processing, abstract processing and spatial perception.[6] The Block Design subtest looked at how an individual is able to analyze and create abstract designs or recreate an already existing design used coloured block pieces.[6] Researchers also used the Digit Symbol and Symbol Search subtest from the WAIS-III to investigate working memory. The last subtest used to analyze fluid intelligence is the Digit Span Backward subtest which comes from the Wechsler Memory Scale-III test.[1] Memory was examined similarly to how fluid intelligence was examined during the study. Researchers used various subtests from the WMS-III to evaluate the multiple aspects of memory. The Logical Memory (delayed and immediate) subtest and the Verbal Paired Associates (delayed and immediate) subtest both helped evaluate the immediate memory of a participant.[7] The Spatial Span (delayed and immediate) subtest was used to assess an individual's working memory.[7] The Digit Span Backward subtest and Letter Number Sequencing subtest from the WAIS-III were also used to help look at short-term memory. Speed of information process was another branch of cognition that researchers in the Lothian Cohort Study wanted to investigate. They did this by using the Symbol

Search and Digit Symbol subtests, both from the WAIS-III.[1] The researchers also evaluated the reaction times and visual inspection times of the participants to gauge how quickly an individual in the study can process information.[1] The Moray House Test was used to evaluate the individual's ability to reason verbally.[1] The test is written out and assesses an individual's general cognition.[1] The Moray House Test was also done on the participants during the first wave of the study in 1947.[1] The scores between the two waves can help provide insight on how each participant's level of general cognition may have changed or shifted. Next, vocabulary and reading was another test that was done on the participants. In order to test for an individual's vocabulary and reading, the National Adult Reading Test was used. It essentially looked at how a participant can pronounce 50 irregular words from the English language.[1] Finally, the verbal fluency of the participants in the study was investigated. To test for verbal fluency, the researchers involved in the study individually asked the participants to say as many words they could think of that started with the letters C, L, and F within 60 seconds for each letter.[1]

Analyzing the Data

During the first wave of the study, childhood intelligence was evaluated for each participant so that data could be compared between a participant's initial level of cognitive functioning and the level of cognitive functioning after several decades. Although childhood intelligence can sometimes be a factor in how an individual's cognition can function when they are older, the researchers conducting the study ensured only to investigate how bilingualism can affect the cognitive ageing of an individual.[1]

Additionally, researchers grouped participants based on how proficiency in their second language, how often they use their second language, and at what age they learned their second language.[1] Grouping participants in such a way helped to control for any confounding variable regarding bilingualism.

Results and Limitations of the Study

The study was able to find many fascinating relationships between slower cognitive ageing and bilingualism. Researchers found that bilingualism does indeed help slow cognitive ageing regardless of the level of intelligence an individual naturally has.[1] The data showed that bilingualism had a consistently positive effect on reading, general intelligence and verbal fluency.[1] This is likely because when someone is bilingual, they tend to have a higher familiarity with a more extensive range of words compared to someone who is monolingual.[1] One of the most critical findings from the study is that bilingualism does not negatively impact intelligence or cognitive functioning at an older age, contrary to what many people thought in the twentieth century.[1] Additionally, some of the participants in the study were multilingual. Researchers found that people who knew three languages had even better cognitive functioning at an older age than participants who knew two languages.[1] This was also such an important finding because there have been studies in the past that argued that multilingual people did not have a higher level of intelligence and cognitive functioning compared to bilingual people.

Although the study setup helped to control for confounding variables, there are some limitations to the study. To determine whether or not someone is bilingual, the participant answered a questionnaire. Thus being categorized as bilingual was determined by questions from a survey instead of language tests and proficiency.[1] Some might feel that the method used to determine bilingualism in the study is an inadequate representation of language proficiency. Another limitation of the study is that researchers were unable to determine if socioeconomic factors or gender caused any influence on cogntive ageing instead of strictly bilingualism.[1] Additionally, the sample size for the study composed of European participants, specifically in Edinburgh.[1] The first language of all the participants were English.[1] Hence, there was not a lot of diversity within the sample that was studied. Nonetheless, the results and insights that the study provided on the effects of bilingualism on cognitive ageing are still relevant and can help pave the way for future studies.

Investigating Dementia and Bilingualism

As some people get older, dementia can become a growing risk. Although ageing is normal and a part of life, dementia is not. Alzheimer's disease and dementia have shown to have correlations to bilingualism. Many recent studies have shown compelling evidence that bilingualism can help serve as a protective measure against the growing risk of dementia.[8] The first study to investigate dementia and bilingualism was published in 2007 by Ellen Bialystok and her colleagues which looked at hospital records of over 200 patients.[8] Researchers found that people who have been blinguals for majority of their lifetime had a four year delay in the initial symptoms of dementia compared to monolinguals.[8] After those findings were discovered, more studies began to start regarding the correlation between bilingualism and the slow of cognitive ageing. Two years after the Bialystok study was published, another case study was developed, which exhibited very similar findings. It was found that patients who were diagnosed with Alzheimer's disease were diagnosed approximately 4.3 years later than monolingual people, and the onset of symptoms for bilingual people did not begin until 5.1 years after monolingual people started to see symptoms.[8] Similar findings have continuously been discovered and will help influence the future of studies regarding bilingualism and the delay of dementia.

Conclusion

Although it was deemed negatively to be bilingual for the majority of the twentieth century, many influential studies have shown in recent years how beneficial bilingualism really is. For the last couple of decades, there has already been a cultural shift that shifted to viewing bilingualism more positively. Thanks to the early studies that discovered findings on bilingualism and cognition, there will undoubtedly be more research and studies conducted on how bilingualism can help slow cognitive ageing, it seems that more and more people will become motivated to learn new languages and develop these skills.[8] While there is still so much that yet to be discovered about the relationship between bilingualism and cognition, the future of these discoveries looks very bright.

Chapter 5: Misconceptions and popular culture

Written by Haya Sonawala

Nearly any topic under the sun can fall victim to myths and misconceptions, which can then be further enforced by popular culture. We live in a world where information is easily accessible at any given time. With the constant development of technology, anyone with access to the internet can have their questions answered in a matter of seconds. This sharing of knowledge can be extremely beneficial, however, it can also be extremely problematic. The internet is a place of endless information, yes, but it is also a place of constantly clashing opinions. This clash is inevitable when any group of people have access to a platform on which they can share their thoughts. As a result, many people may not have their questions answered or they might be exposed to outdated information or information that isn't credible, which they mistakenly take to be true. This spread of misinformation and misconceptions can be especially powerful when individuals are looking for information on topics that are foreign to them.

Popular culture can be defined as "the set of practices, beliefs, and objects that embody the most broadly shared meanings of a social system."[1] This can include "media objects, entertainment and … linguistic conventions."[1] This chapter will take a look at how bilingualism and cognitive ageing are presented on these forums, and whether or not these depictions enforce or oppose existing misconceptions regarding these topics.

Bilingualism and Two Worlds: Xenophobia, Elitism, and Popular Culture

In the book *Popular Culture, New Media and Digital Literacy in Early Childhood*, Charmian Kenner highlights two common assumptions regarding bilingual children that live in English-speaking countries. The first of these assumptions is that they are thought to be "living in 'two worlds',"[2] one based around each respective language. The other is that these "children may experience difficulty, or even trauma, in connecting these two worlds."[2] There is a popular Nelson Mandela quote that enforces this idea of two worlds, or at least, the idea that each language represents two distinctly different aspects of an individual: "If you talk to a man in a language he understands, that goes to his head. If you talk to him in his own language, that goes to his heart."[3] This widely known quote, however, is actually a misquote. The original quote is "when you speak a language, English, well many people understand you, including Afrikaners, but when you speak Afrikaans, you know you go straight to their hearts."[3] An important distinction should be made between these two quotes. Firstly, the former may be classified as popular culture. The original quote was altered in a way that generalized it, making it more relatable to the masses; it should be noted that popular culture is very closely linked with mass culture.[1] Mass culture can also be referred to as the culture of the majority. The second notable distinction to be made between these two quotes is that "Mandela was talking about Afrikaans, the language of the oppressor," and he was saying that "South Africans should learn Afrikaans as part of the fight against Apartheid."[3] The context provided by this distinction introduces an entirely new dynamic between two languages—that of the oppressor and the oppressed.

In English-speaking countries the language of the oppressor is English. In these circumstances, however, the oppressed are not battling Apartheid, rather, they are battling everyday racism and xenophobia. The phrase "you're in America, speak English" or some variation is widely known and heard. The sentiment behind the phrase would surprise almost no one who is bilingual or is familiar with a bilingual person in a predominantly white and English-speaking country, particularly if the bilingual individual's second language is that of an ethnic minority. Speaking the language of the oppressor has a very different effect in each case. As Mandela suggested, it can be part of the fight against the Apartheid, however in the case of an English-speaking country, speaking the language of the oppressor is distancing one's self from their minority culture and in turn enforcing the 'otherness' that can often cause doubt and lead to further misinformation. The assumption that the two worlds associated with the two languages leads to trauma further perpetuates the aforementioned 'otherness', as well as the idea that bilingualism, specifically when there is a different cultural upbringing associated with each language, is a bad thing for kids. This is particularly harmful because most bilingual children in English-speaking countries are children of immigrants, and "the value ascribed to English as the dominant language marginalises [their] linguistic communities."[2] The assumption suggests that in order for a child to be mentally healthy they must assimilate in one over the other; given that these assumptions are regarding children in an English-speaking country, it can be understood that children would be expected to better assimilate in the English-speaking culture. This expectation risks a loss of not only minority languages but also cultures over generations in English-speaking countries, because children who do not learn their culture will not be able to pass it on to their children. "Children are actors in the remaking of culture," so it is important to provide them with the resources and experiences that will allow them to create a fusion culture that is well balanced.[2]

This cultural erasure, however, is not always the case in popular culture, though it is a popular assumption that comes with globalisation. With new technology and globalization, the melding of different cultures in popular culture is much more prominent, which "show[s] the changing nature of cultural practices and inter-relationships ... [and] while such material could be seen as the colonisation of indigenous culture by Anglo-American global forces, the results are undeniably complex."[2] A theory developed by George Ritzer suggests the idea of the McDonaldization of society.[4] It stems from "the process by which the principles of the fast-food restaurant are coming to dominate more and more sectors of American society as well as of the rest of the world."[4] This idea has been extrapolated to apply to American culture as a whole, and how it has settled into aspects of many other cultures. This is applied to children's popular culture, which "is often considered to be subject to the 'McDonalidisation effect' in that it is assumed that US-based themes dominate the global market."[2]

Kenner challenges the aforementioned assumptions—the first being that bilingual children in English-speaking countries live in two worlds, and the second being that the reconciliation of these two worlds can be traumatic for the children—as well as the idea that children are more exposed to English culture due to the popular culture to which they are exposed. In doing this they prove that the assumptions are not only largely problematic, but also untrue. Kenner then goes one step further, and addresses the importance of representation in popular culture, particularly for bilingual children. Counter to the assumptions that bilingual children have two worlds that can be traumatic to reconcile, research focusing on bilingual children of ethnic minorities in London found "first that their cultural worlds are hybridised rather than separate, and secondly that they create further hybridity through the making of texts which represent their complex cultural identities."[2] Children might encounter media and texts in their non-English language, but the source of this exposure is limited to their families or their cultural communities: "children's cultural experiences in languages other than English are relatively little known."[2] This exposure is also subject to debate, because, as mentioned before, the 'McDonaldization' of popular culture jeopardizes the cultural integrity of the non-English shows, movies, music, etc. that bilingual children to which bilingual children are exposed. Children will, however largely "become familiar with mainstream popular culture in English via the media and peer group interaction at primary school."[2]

Kenner does not deny that Anglo culture is dominant in TV and other forms of popular cultures, and acknowledges that there are "TV programmes directly modelled on ones of Anglo origin" and that certain movies or videos are much more Westernized now, so much so that they are "far removed from the Western stereotype[s]" of those cultures that enforce a foreign and unfamiliar image rather than the familiar aspects that Anglo people are finding in foreign popular culture.[2] While these Anglo-rooted conventions are widespread in non-English popular culture, other cultures have also influenced English popular culture. A very notable example of this is "dramatic Mexican-style soap operas … [that] can occasionally have some effect on mainstream Anglo culture."[2] Take the popular show "Jane the Virgin" as an example. The TV show is centred around a Venezuelan family and the show itself is classified as an "American telenovela,"[5] heavily drawing on conventions of the popular South American genre. The show even has a character that is a telenovela actor and a main character that almost exclusively speaks in Spanish, thus exposing its largely English-speaking audience to not only South American culture but also to another language. It should be noted that while Spanish is often taught in American schools, Spanish speakers, particularly from South America, are not always well respected, as is addressed on the show. Kenner also refers to the influence of other cultures on mainstream Anglo popular cultures seen in the "recent interest in the traits of Bollywood productions and Eastern martial arts films."[2] In this way, bilingual children are exposed to the worlds that are ascribed to each of their respective languages. These worlds, often already being merged into one, challenge the assumptions that bilingual children live in two distinct worlds, and that merging them is traumatic for them, since they are already seeing the two melded together in a lot of mainstream popular culture.

Cognitive Ageing and Globalization in Popular Culture

Young children are very impressionable, and they are significantly molded by the people around them as well as the media to which they are exposed. Many people, particularly in the west, have a "cultural bias in favor of youth,"[6] and this is further perpetuated as children develop this bias. As a result, "it does not come as a surprise that children already develop old-age prejudices at a very young age."[6] The exposure that children have in their everyday life tend to be child and youth oriented,[6] and as such they do not have a lot of experience with elderly people, with the exception of their own grandparents.

Oftentimes, the forms of popular culture that are aimed towards children either have very little representation of older people, or they perpetuate harmful stereotypes about growing old and older people. These mediums often emphasize "negative characteristics such as physical unattractiveness and passiveness."[6] The theme of ageing is hardly ever central in popular media aimed towards children, however there are two Dutch children's books that focus on realistic representations of ageing. The representation of the elderly shown in the books, however, is actually quite damaging for children. Both books "discuss the right to self-determination, in this case the right of older people to die when they feel ready, either because they see their life as completed or because they suffer from old age or want to prevent future misery."[6] This idea suggests that after one reaches a certain age there would be nothing in their life besides misery and sadness, and that the elderly cannot do anything productive or useful late in their life. It dehumanizes them in that it suggests elderly people are burdens and not people with their own sets of aspirations of their own.

As previously stated, however, the old-age prejudice is something fairly specific to Western cultures. Elderly people are central to many Eastern cultures, and as a result have a much more prominent role in their popular culture. An example of this is Japanese fairy tales, which "abound with aged heroes and heroines."[6] While Western fairy tales do feature old people, those individuals often take the role of supporting characters or, even worse, of villains. The evil queen from Snow White is a perfect example of this—she is young and beautiful when she is introduced, but when she actually takes actions against the story's protagonist, she is old. Fairy tales are often didactic stories for children, and the story of Snow White inadvertently teaches them to be afraid of the elderly. In Japanese fairy tales, however, the protagonist is more often than not old people who "undergo adventures and perform difficult tasks,"[6] and as such children are taught to admire them and actually hope to be like them one day.

The benefits of the globalisation of media and popular culture were previously discussed in the context of bilingualism and combating assumptions regarding bilingual children. This is another case in which the globalisation of popular culture can be extremely beneficial. This is particularly the case because of how concentrated the assumption about cognitive ageing is. By learning a new language or by being exposed to translated texts, monolingual English-speaking children can experience positive representation of cognitive ageing and view their own old age as a time of opportunity rather than a time of fear. All languages are associated with their respective culture(s), and we can either look at those languages as barriers or differences that we should be afraid of and avoid, or we can view them as opportunities to challenge our prejudices and practice empathy.

References

Ronjat, J. (1913). Le development du language observe chez un enfant bilingue. Paris: Champion.

Crnac, A. (2019). Bilingualism and Cognition: The role of early bilingualism in the development of executive control and its influence on cognitive abilities. http://hdl.handle.net/1946/32203

Bialystok, E., Craik, F. I. M., & Freedman, M. (2007). Bilingualism as a protection against the onset of symptoms of dementia. Neuropsychologia, 45, 459-464.

Saer, D. J. (1923).The effects of bilingualism on intelligence. British Journal of Psychology, 14, 25-3

McCarthy, D. A. (1930). The language development of the pre-school child.- Minneapolis: University of Minnesota Press.

Adesope, O., Lavin, T., Thompson, T., & Ungerleider, C. (2010). A Systematic Review and Meta-Analysis of the Cognitive Correlates of Bilingualism. Review of Educational Research, 80(2), 207-245. Retrieved August 17, 2021, from http://www.jstor.org/stable/40658462

Pintner, R. (1932). The Influence of Language Background on Intelligence Tests. The Journal of Social Psychology, 3(2), 235-240. doi: 10.1080/00224545.1932.9919147

Fukuda, T. (1925). A survey of the intelligence and environment of school children. American Journal of Psychology, 36, 124-139

Brunner, E. D. (1929). Immigrant farmers and their children. New York: Doubleday, Doran, & Co.

Arsenian, S. (1937). Bilingualism and mental development. New York: Columbia University Press.

McLaughlin, B. (1978). Second language acquisition in childhood. Hillsdale, N.J.: Lawrence Earlbaum Associates.

Cummins, J. (1976). The influence of bilingualism on cognitive growth: A synthesis of research findings and explanatory hypotheses. Working Papers on Bilingualism, 9, 1-43

Peal, Elizabeth, & Lambert, Wallace E. (1962). The relation of bilingualism to intelligence. Psychological Monographs, 76(27, Whole No. 546).

Carrow, S. M. A. (1957). Linguistic functioning of bilingual and monolingual children. Journal of Speech and Hearing Disorders, 22, 371-3

Diaz, R. (1983). Thought and Two Languages: The Impact of Bilingualism on Cognitive Development. *Review of Research in Education, 10*, 23-54. doi:10.2307/1167134

Bialystok, E. (1987). Words as things: Development of word concept by bilingual children. Studies in Second Language Acquisition, 9, 133-140.

Bialystok, E. (2001). Against isolationism: Cognitive perspectives on second language research. Selected Proceedings of the Second Language Research Forum, 97-103.

Leopold, W. F. (1939). Speech development of a bilingual child: A linguist's record (4 vols.). Evanston, Ill.: Northwestern University Press.

Cazden, C. R. (1974). Play with language and metalinguistic awareness: One dimen- sion of language experience. Urban Review, 7, 28-29.

Harris, C. W. (1948). An exploration of language skill patterns. Journal of Educational Psychology, 32, 35

Bialystok, E. (2005). Consequences of bilingualism for cognitive development. In J. F. Kroll & A. M. B. de Groot (Eds.), Handbook of bilingualism: Psycholinguistic approaches (pp. 417-432). New York, NY: Oxford University Press.

Bialystok, E., Craik, F. I. M., & Freedman, M. (2007). Bilingualism as a protection against the onset of symptoms of dementia. Neuropsychologia, 45, 459-464.

Bialystok, E., & Martin, M.M. (2004). Attention and inhibition in bilingual children: Evidence from the dimensional change card sort task. Developmental Science, 7, 325-339.

Understanding the Dynamics of the Aging Process [Internet]. National Institute on Aging. [cited 2021 Aug 23]. Available from: http://www.nia.nih.gov/about/aging-strategic-directions-research/ understanding-dynamics-aging

The cooperative human. Nat Hum Behav. 2018 Jul;2(7):427–8.

Yorkston KM, Bourgeois MS, Baylor CR. Communication and Aging. Phys Med Rehabil Clin N Am. 2010 May;21(2):309–19.

Sirbu A. The significance of language as a tool of communication. In Constanta, Romania; 2015.

Steber S, Rossi S. The challenge of learning a new language in adulthood: Evidence from a multi-methodological neuroscientific approach. PLOS ONE. 2021 Feb 19;16(2):e0246421.

Pearson BZ. Raising a Bilingual Child. Diversified Publishing; 2008. 353 p.

Byers-Heinlein K, Lew-Williams C. Bilingualism in the Early Years: What the Science Says. Learn Landsc. 2013;7(1):95–112.

Sirbu A. Language interference triggered by bilingualism. In Constanta, Romania; 2015. Available from: https://www. researchgate.net/publication/337472381_LANGUAGE_INTERFERENCE_TRIGGERED_BY_BILINGUALISM

D'Acierno MR. Three Types of Bilingualism [Internet]. 1990 [cited 2021 Aug 23]. 65 p. Available from: https://eric. ed.gov/?id=ED321574

Diller KC. "Compound" and "Coordinate" Bilingualism: A Conceptual Artifact. WORD. 1970 Aug 1;26(2):254–61.

Yow WQ, Markman EM. Bilingualism and children's use of paralinguistic cues to interpret emotion in speech. Bilingualism: Language and Cognition. 2011 Oct;14(4):562–9.

Akhtar N, Menjivar JA. Cognitive and linguistic correlates of early exposure to more than one language. Adv Child Dev Behav. 2012;42:41–78.

Bialystok E, Shapero D. Ambiguous benefits: the effect of bilingualism on reversing ambiguous figures. Dev Sci. 2005 Nov;8(6):595–604.

Brito N, Barr R. Influence of bilingualism on memory generalization during infancy. Dev Sci. 2012 Nov;15(6):812–6.

Klimova B. Learning a Foreign Language: A Review on Recent Findings About Its Effect on the Enhancement of Cognitive Functions Among Healthy Older Individuals. Front Hum Neurosci. 2018 Jul 30;12:305.

Marian V, Shook A. The Cognitive Benefits of Being Bilingual. Cerebrum. 2012 Oct 31;2012:13.

Pliatsikas C, Meteyard L, Veríssimo J, DeLuca V, Shattuck K, Ullman MT. The effect of bilingualism on brain development from early childhood to young adulthood. Brain Struct Funct. 2020;225(7):2131–52.

Hewings-Martin Y. Bilingualism: What happens in the brain? [Internet]. Medical News Today. 2017 [cited 2021 Aug 24]. Available from: https://www.medicalnewstoday.com/articles/319642

Internal Administrator. The cognitive benefits of being. Dana Foundation [Internet]. 2012 [cited 2021 Aug 24]; Available from: https://dana.org/article/the-cognitive-benefits-of-being-bilingual/

Bilingual effects in the brain [Internet]. Nih.gov. 2015 [cited 2021 Aug 24]. Available from: https://www.nih.gov/news-events/nih-research-matters/bilingual-effects-brain

Bak TH, Nissan JJ, Allerhand MM, Deary IJ. Does bilingualism influence cognitive ageing? Annals of Neurology. 2014;75(6):959–63.

The Scottish MENTAL SURVEY 1947 [Internet]. The Scottish Mental Survey 1947 | The Lothian Birth Cohorts of 1921 and 1936. [cited 2021Aug24]. Available from: https://www.lothianbirthcohort.ed.ac.uk/content/scottish-mental-survey-1947#:~:text=The%20Scottish%20Mental%20Survey%20 1947%20was%20carried%20out%20by%20the,schools%20 on%20June%204%2C%201947.

Mattschey J. Repeating the same mistakes: A century of research interested in the effects of bilingualism on non-linguistic cognitive functions. 2020;

Cherry K. What are fluid intelligence and crystallized intelligence? [Internet]. Verywell Mind. 2021 [cited 2021Aug24]. Available from: https://www.verywellmind.com/fluid-intelligence-vs-crystallized-intelligence-2795004

Cherry K. How does the Wechsler adult intelligence scale measure intelligence? [Internet]. Verywell Mind. 2020 [cited 2021Aug24]. Available from: https://www.verywellmind.com/the-wechsler-adult-intelligence-scale-2795283

Wechsler IQ Test Subtests and Suggestions [Internet]. Books to enrich thinking and learning. Think tonight; 2021 [cited 2021Aug24]. Available from: https://www.thinktonight.com/WISC_IV_subtests_s/331.htm

Stebbins GT. Neuropsychological testing. Textbook of Clinical Neurology. 2007;:539–57.

Antoniou M, Wright SM. Uncovering the mechanisms responsible for why language learning may promote healthy cognitive ageing [Internet]. Frontiers. Frontiers; 1AD [cited 2021Aug24]. Available from: https://www.frontiersin.org/articles/10.3389/fpsyg.2017.02217/full

Popular Culture. (n.d.). Oxfordbibliographies.Com. Retrieved August 23, 2021, from https://www.oxfordbibliographies.com/view/document/obo-9780199756384/obo-9780199756384-0193.xml

Jackie, M. (Ed.). (2004). *Popular culture, new media and digital literacy in early childhood.*

Taylor & Francis Group.*My favorite Nelson Mandela (mis)quote.* (n.d.). Harvard.Edu. Retrieved August 24, 2021, from https://scholar.harvard.edu/pierredegalbert/node/632263

Ritzer, G. F. (2012). *The McDonaldization of society: 20Th anniversary edition* (7th ed.). SAGE Publications.

(N.d.). Wikipedia.Org. Retrieved August 25, 2021, from https://en.wikipedia.org/wiki/Jane_the_Virgin

Wiener, R. L., & Willborn, S. L. (Eds.). (2014). *Disability and aging discrimination: Perspectives in law and psychology* (2011th ed.). Springer.

Joosen, V. (Ed.). (2018). *Connecting childhood and old age in popular media.* University Press of Mississippi.

www.ingramcontent.com/pod-product-compliance
Lightning Source LLC
Chambersburg PA
CBHW021544270326
41930CB00008B/1363